The Individual in Darwin's World

STEPHEN JAY GOULD

THE SECOND
EDINBURGH MEDAL
ADDRESS

Set in Digitek Janson by
Waverley Graphics Ltd., Edinburgh
and printed in Great Britain by
Ritchie, Edinburgh and
bound by Hunter and Foulis, Edinburgh

ISBN 0 7486 0227 5

The Second Edinburgh Medal Address transcribed and edited by Ian Wall, Ian Rolfe and Simon Gage of the City of Edinburgh District Council, and Vivian Bone and Helen Simms of Edinburgh University Press.

PROFESSOR STEPHEN JAY GOULD

This is a transcript of an oral presentation delivered without written text. Please remember that written and spoken English are different languages, and that the latter does not translate well into the former mode.

Stephen Jay Gould
Professor of Geology
Museum of Comparative Zoology
Harvard University

The Lecture

In the city that produced our greatest theory on the efficacy of individual action in history and social order, namely the laissez-faire economics of Adam Smith to which I'll return later; I thought that the subject of the role of individuals in the world that I know best, namely Darwin's world, might be welcomed, and so the title of this address is 'The Individual in Darwin's world'. You also realise of course that there is a connection between Edinburgh and Darwin. Not the happiest connnection of his life – he studied medicine here and liked neither medicine nor, I think, everything about Edinburgh, particularly the lack of light and the cold of December. But nonetheless there is a Darwinian connection. It may be more important than we realise, because his first evolutionary ideas did not come from the Galapagos. They may well have come from around here, according to Adrian Desmond's latest book.

I offer two disclaimers about this talk. First, I don't think I've ever given a talk without slides before: primates are visual creatures. Secondly, I am juxtaposing two very disparate subjects in talking about the individual in Darwin's world: in fact the title is a con – or rather a double entendre – although people do say I deal with connections.

The first topic – maybe they are quite separate – is our status, the status of us as individual human beings; our status as evolution suggests with respect to our hopes of what our culture has held out for – and that is the individual, namely us, in Darwin's world – not the world of Darwin's theory but the factual, actual world of the history of life.

And then secondly, a quite different theme, the seemingly abstract and surprisingly relevant subject of what new developments in evolutionary theory mean for the concept of individuality itself in the broadest sense, to the centrality of personhood. And that is the individual, in the largest sense of the term in Darwin's intellectual world this time, within the theory, whereas the first part is within our attempt to understand the actual pattern of life's history.

The common theme to this talk, 'The Individual in Darwin's World', these two very disparate parts, is that our assumptions about the nature of the world, the way we think the world actually works, so often record our hopes about the world and not the nature of Nature. Now this carries a superficially depressing message because our hopes do get punctured by the actual quality of life and its history and the theory that we use to understand it. I'd like to suggest that it is quite unnecessary to see them in a pessimistic light. One part will argue that human evolution was not destined to be and that we're lucky to be here at all and I think that is perhaps superficially depressing. The second part will argue that personhood or individuality has no special status, and I suppose that can be seen as somewhat depressing, but I do not mean it that way and I will try to make that clear.

So I move into the first part which is concerned with the individual, namely our status as human beings, within the world of life's history, the evolutionary world that Darwin spelled out for us. It is quite interesting if you look at Darwin's theory and grasp the subtleness – you realise that Darwin almost uniquely among nineteenth-century evolutionists did not preach a notion of intrinsic progress as part of the workings of evolution itself. He is quite clear on that. Late in his life Darwin writes to Hyatt, an American paleontologist whose office I now occupy at Harvard, and who was a scientist who thought that there was a necessary advance leading up to human evolution. Darwin writes, 'After long reflection I can not avoid the conclusion that no inherent tendency to progressive development exists.' Nonetheless, I think it is quite clear that the most common misunderstanding of evolution, at least in lay culture, at least in the United States, is the equation of evolution with some notion not only of progress but of our eventual appearance as somehow intrinsically inevitable and predictable within the confines of the theory. I put as the best evidence of this, at least from an American standpoint, a very interesting survey that recently appeared. Five-hundred newspaper editors, a very large sample of opinion-formers of the United States, people with no problem with evolution as a general notion, were asked as one of the questions of the survey to check one out of five boxes representing what they thought was the best characterisation of evolution: Only about 20 out of 500 checked what was in fact the best answer, namely that evolution was differential reproductive success of some organisms versus others. There were two others which were just wrong and not many people checked those. But almost everyone checked one of the two following: 160, a good number but not a majority, checked that evolution meant the survival of the fittest – a somewhat truncated notion but not totally incorrect. But 260, virtually an absolute majority and a hundred more than any other answer, checked the worst answer of all, which stated evolution means the inevitable predictable progress of life from monad to human being along predictable pathways. So that clearly is, at least in popular culture, a very common misunderstanding. Now we ask why it's such a common misunderstanding. I really think it is a question of our culture and our hopes for solace in the world we know. But at least from the standpoint of a geologist, my profession, I think it is a direct result of what I like to call geology's most frightening fact. Geology discovered that time comes not in thousands but in millions and billions of years. That in itself is not necessarily a threat to human views of our own intrinsic importance. Thomas Browne in his 'Religio Medici' in the mid-seventeenth century wrote a beautiful epigram of the older, comforting view: 'Time we may comprehend. 'Tis but five days elder than ourselves.' which is a comforting view. Then comes along James Hutton in volume one of the *Transactions of the Royal Society of Edinburgh* and there's 'No vestige of a beginning and no prospect of an end': time cycles through endless amounts. Now that's not in itself a threat. What is a threat is that time comes in millions and billions of years *and* human history is restricted to the last millimicrosecond. If human history still pervaded all those millions and billions you could say that we were intrinsic and predictable. But that's not so. Time comes in billions and we're at the very end. We occupy such a tiny amount of the total structure of time that pedagogues throughout the world can only discuss it in metaphor, so that we talk about the last inch of the cosmic mile or the last minute of the cosmic year. My favourite metaphor is by the great American writer John McPhee; it says: consider the history of the Earth to be as long as the old measure of the English yard which is the distance from the King's nose to the tip of his outstretched finger, and as McPhee says, one stroke of a nail file on the third finger would wipe out human history.

Now that's frightening because the very existence of that fact forces one to come to grips with the terrible consequence, which I think is quite true, that maybe we were not destined to be. Maybe we really are a cosmic afterthought, a tiny little twig on this arborescent tree of life, a little side branch that would never appear again if we run the tape again. That is geology's frightening fact and I think it is the main reason from the geological standpoint why we feel impelled to tell the story of life's history as a tale of predictable progress leading up to us. We can say, 'Yes, we're the last millimicrosecond, but everything that came before is predictably preparatory for the necessary and eventual appearance of human beings, and it but doesn't matter that we're the last millimicrosecond because everything's part of the same story and must go in that direction'.

Now the best argument against that was presented not by a scientist but by America's greatest humorist, Mark Twain, who saw clearly, as he usually did, and who wrote late in the last century, when the Eiffel Tower was the world's tallest structure: 'Man has been here thiry-two thousand years. That it took a hundred million years to prepare the world for him is proof that that is what it was done for. I suppose it is, I dunno. If the Eiffel Tower were now representing the world's age, the skin of paint on the pinnacle knob at its summit would represent man's share of that age; and anybody would perceive that the skin was what the tower was built for.' Nonetheless, despite Twain's warning, you can not get away from that desire to see the history of life that way. And so there are two separate subjects which we read in the light of our hopes as a predictably, gradually progressive story leading to us. First is the story of increase in numbers. We start with a few simple microbial things 'way back when', you get more and better until you come up to a world full of algae, hippos, flamingos, petunias and people. And then there's the story of increasing the complexity of form – you start with that same microbe, and things, or at least the most complex thing, gets more and more complex through time until we finally reach human consciousness that microsecond ago. Now, in response to that, which I think is utterly wrong as a way of looking at the history of life, I think in fact the plain language implication of geology's most frightening fact, namely that if we're that last millimicrosecond we did not have to be, we are a piece of luck, is fundamentally correct, and it is in that sense that we ought to read this whole first part of my talk. We have got to refocus and reconstitute our view of the individual regarding this world, the history of life, and the individual. The thought experiment that I would like to use in discussing this is that of replaying life's tape, if you will. What would happen if we could rewind the tape of life to some moment in the fairly distant past and simply let it run again? Would we get fundamentally the same result, with little changes here and there to be sure? I would like to suggest to you that you should seriously entertain the notion that what you would get would be totally different. Quite unpredictably different. You would get something and you would be able to make sense of it after the facts, but you would never get the same thing twice or even close to it. Now when I mention a theme like that I think that what it raises is a fear of the chaotic, the random in the vernacular sense, meaning something utterly without any possible explanation. That is really not what I mean at all. What I am talking about is the theme that historians call 'contingency', the notion that given the inordinate complexities of the thousands of unpredictable steps inherent in any complex historical story, and given the large dollop of randomness and irreducible complexity that does exist in the world, you would never get the same thing twice. Whatever happens makes sense after the facts, it is not random and chaotic but it is utterly unpredictable before. If you could rewind the tape back to the origins of multicellular life six-hundred milllion years ago you would never know what to predict nor would you ever get again anything like what we got. And since there is only one little lineage that ever came close to consciousness, namely us, and no other lineage on Earth shows any threat of moving in that direction I rather suspect that most replays would not include the evolution of any creature like us.

Now this theme of contingency is very much beloved of literature more than of science and I think if you want your analogues for it you look towards literature rather than to science. It is the kingdom lost for want of a horseshoe nail. It is the fact that Heathcliff left that conversation between Cathy and Nelly Dean just a few minutes before he should have. If he'd only sat and listened he would have spared himself a lot of turmoil but it would have been a worse book. Or you read the various prefaces to *War and Peace* on Tolstoy's obsession with that issue of the role of the individual versus the role of inexorable forces in history and how they interact. Or if you like modern literature you read Kurt Vonnegut's *Galapagos.* Kurt Vonnegut decided to go to the Galapagos as a prelude to writing a novel about evolution. The theme he chose was not natural selection, or any one of a hundred themes he could have chosen, but historical contingency – the notion that what happens is so complex, even though sensible, that it could never occur in the same way again, and you get a different result. His novel *Galapagos* is not only about that, it is about how all of human history through a weird set of circumstances gets concatenated into an oddly assembled group of people on a boat sailing off the Galapagos. Or you go to the movies. You look at parts one and two of 'Back to the Future' which are films about contingency. Marty McFly's struggle to make his past come out as *his* past so that he can continue to exist. Or you look at the greatest American classic of that genre, 'It's a Wonderful Life' – that's why I called my recent book on this subject *Wonderful Life* – and you look at the famous scene of George Bailey about to kill himself in despair, being shown by his Guardian Angel that his life did make a difference, that without him and his apparently insignificant life the history of his town would have been utterly different and much, much worse. The contingency is there but does it apply to the history of life? I cannot give you in the time available much of a feel about it, but let me just discuss two subjects under this theme of unpredictability and non-repetition. First of all on the question of increasing complexity, you know people will say when I talk about this subject, 'Look, you're not being fair, you're worrying something to death. Surely you don't deny that early in life's history there were just single celled creatures on a bacterial level of organisation and now there are all these very complex things. So surely there must be a predictable trend in increasing complexity.' I do not want to be a curmudgeon, yet I would urge a different view about the nature of complexity and the history of life. While not denying that the most complex creature on Earth has tended to be more complex through time, we are really looking at it in an odd way because it is more an expansion from the necessary simplicity of beginnings rather than a necessary predictable drive towards higher states of organisation. Life for chemical reasons originated right at the lower bound of its conceivable preservable complexity. You cannot begin with a lion; you begin with something really simple, some chemical precusor. And if you begin there, there is no room below it for less complexity. If life is going to expand it's going to occasionally add more complexity, and there is a drive in that direction; it is just a general expansion. Every once in a while, at the only end open, you add something that is more complex, but the modal organism on Earth, the most common organism on Earth, remains a simple prokaryotic cell, a bacterial level of organisation. That has never changed in the history of life on Earth. It is as true now as it was when the first living creatures arose. There are more *E.coli* bacteria in the digestive tract of every one in this room than there have ever been human beings on Earth. And those bacteria are going to outlive us; they cannot be nuked to oblivion. The modal organism on Earth has never changed. So we look at complexity the wrong way because we are complex and we tend to see a drive in that direction. It is not; it is just an occasional expansion, asymmetric expansion into that domain. But that is not the main thing I want to discuss; what I want to discuss is that contingency of the replay thought experiment. Again people say to me, 'But you are being a curmudgeon because it is not fair, again you start with single creatures, you end up with human beings and there has got to be increased complexity in between, so it must be predictable.' I do not know. If it was that kind of story then tell me why about five sixths of life's history is only the history of single celled creatures. If there was such a great drive, and advantage to that complexification, why did it take so long? I know there are answers based on different atmospheric conditions, I do not know whether they are right but I think it is curious that if there is a necessary predictable drive in that direction, that it took so long happening. Nonetheless if when it happens you have then got a sequential condition of greater complexity through the last 600 million years I might be more willing again to allow that theme. But that is not true either. In fact almost all the major anatomical designs of organisms appear in one great whoosh called the Cambrian Explosion about 600 million years ago. You realise that a whoosh or an explosion in geological terms has a very long fuse. It can take a couple of million years, but a couple of million years in the line of billions is nothing. And that is not what that world of necessary, predictable advance ought to look like. You should not get five sixths of the history of life as single cell creatures and then essentially all the designs of multicellular life appearing in one incident, and then 600 million subsequent years which in one very major sense can be seen as little more than variations on a theme. So let's talk for a moment about that cardinal event, namely the Cambrian Explosion. It is not until recently that we have had any sense of the full range of diversity that the Earth contained, coincident with and right after this Cambrian Explosion. To get some sense of the full range of diversity you have to have the good fortune of finding one of these rather rare faunas of soft bodied creatures in the fossil record. Most of the fossil record is the story of hard parts, and hard parts are relatively uninformative for two reasons: first of all a lot of creatures do not have hard parts, so you are missing more than half the kinds of organisms that only have hard parts; secondly a hard part often does not tell you much about the creature. If all you had was clam shells and all clams were dead you would not know much about the anatomy of the animal inside. Lucky geological circumstances conspire every once in a while to permit the preservation of soft bodied fossils. Luckily we have such a fauna from 'right after' the Cambrian Explosion (a couple of tens of million years is right after to a geologist) and therefore we have this key fauna from the most crucial time. We want to know if contingency rules the history of life; what the full range of possibilities was when multicellular life first arose. This fauna is called the Burgess Shale, it sits in Yoho National Park on the border between Alberta and British Columbia in Canada. It's been known for a long time; it was discovered by a great American paleontologist in 1909 but thoroughly misinterpreted by this man, C.D. Walcott, who because he was such a conservative, traditionalist thinker felt impelled to interpret the animals of the Burgess Shale in such a way that he could read the history of life in the conventional manner, that is a sop to our hopes and solaces, namely as predictable, progressive increase. So Walcott interpreted all these animals in the Burgess Shale as simple, primitive precursors in modern groups. That is simple, primitive precursors of creatures that came along later. And the world did not know about the Burgess Shale because he interpreted it in such a conventional way that it taught us nothing different about life's history. In the last twenty or thirty years, three paleontologists from this part of the world, Harry Whittington, Simon Conway-Morris and Derek Briggs, have completely reinterpreted the Burgess Shale in a magnificent series of highly technical anatomical mongraphs which have given us an entirely different view of life which I can only summarise breathlessly in a couple of minutes. But what they have shown us is that the Burgess Shale does not just contain a few primitive simple precursors of things that came later. Actually in that single quarry in British Columbia there are more kinds of organisms in terms of basic anatomical design than inhabit the world's oceans today. This indicated that the maximal number of possibilities was right at the start and not slow steady progressive predictable addition later.

For example, this is not the world of mammals or the Age of Mammals that text books would like; it is the world of insects, the world of arthropods, you know that, 80 per cent of animal species are arthropods, probably more, and yet all of them fit into three basic anatomical groups: there is the insect group, there is the spider and scorpion group and there is the crustacean group. And all these millions of species, with great diversity in their ecological role and size and shape but not in terms of basic anatomical plans, go comfortably into one of those groups, with clear and precise basic anatomical definitions. In the Burgess Shale, one quarry in British Columbia, there are some twenty alternative designs for arthropods which did not make it. In the Burgess Shale, and this is even more exciting, there are some fifteen to twenty kinds of organisms so bizarre that we don't even know how to classify them. They just are what they are. One called *Opabinia* has five eyes and a vacuum cleaner-like nozzle coming off the front, which bends around to a mouth below. Another called *Anomalocaris* has a mouth that looks like a camera shutter – it does not work on the hinged principle as almost all modern mouths do. And it is a long creature with an arthropod-like pair of appendages in front. There is another creature so bizarre that its technical scientific name is *Hallucigenia.* Most of these died. The history of life is not a history of expansion and perennial progress; it is a story of decimation, of the loss of most of the initial possibilities. Now you may admit that and still say, 'But so what? There's a perfectly conventional interpretation of that which still allows us our hope of progresss and predictability'. Sure, you can say this is very interesting, I did not know that, but it does not threaten our solace, because after all, all right, there were a hundred and only ten survived. Clearly the ten that survived were those who were anatomically superior, ecologically better adapted. There must have been reasons. So even though there are a hundred possibilities, you could have been a paleontologist back then and said, 'These 10 are going to make it, probably these 90 won't. Now I cannot swear to you that this is false because it is only negative evidence but I can say that Whittington, Briggs and Conway-Morris, all of whom began with that assumption, after working on this fauna for twenty years, could find nothing about those that did not make it that in any way distinguished them or made them predictably less likely to pass through this set of extinctions that must have occurred after the time of the Burgess Shale. So that I think we must seriously entertain the hyphothesis that there is a very strong probability that the decimation of the Burgess Shale fauna is a grand scale lottery, meaning it would not happen the same way again if you replayed the tape and decimated it a second time in different ways. Now if you remember how to calculate permutations and combinations from a high school algebra course – you do not have to remember any more because there is a button on your calculator that does it now – there are how many ways, do you think, to take different groups of 10 out of a pool of 100? And that might be about the right order of survivorship of the Burgess Shale. The answer is 17 trillion. There are 17 trillion different groups of 10 in a lottery model which you can get out of a pool of 100.

Now you might say to me, 'So what? I mean I don't care about invertebrates anyway, so we get a different set'. But I have got news for you. There is an animal in the Burgess Shale which everybody ought to know, the one called *Pikaia*. And the reason you ought to know *Pikaia* is that *Pikaia* is a chordate, that is it is a member of our phylum. It is the first member of our phylum. I am not saying it is our direct ancestor, I am not saying it is the only chordate alive at that time, but believe me it does not look as though chordates were a conspicuously successful group. If you decimate the Burgess Shale differently most of the different decimations do not include the survival of *Pikaia* or any chordate, and we are wiped out of future history. So you may thank your lucky stars for the survival of *Pikaia* through whatever happened to the Burgess fauna, because otherwise this lecture would not have happened.

Now I will give one further example then we will get onto part two. You might say to me, 'Fine. I didn't know that about the Burgess Shale, I didn't know that about *Pikaia, Pikaia's* precious, I will grant you that. Decimate the Burgess Shale differently, you don't get vertebrates. But surely once you get vertebrates, then we are home and free, right?

Everybody knows: fish, amphibians, reptiles, mammals, humans – what used to be called Man – but that is only half of us. I mean, everybody knows that sequence, and it must be so'. But that is not so either. There are hundreds of equally odd contingencies involved in just the lineage of vertebrates. The lineage of fishes that gave rise to terrestrial vertebrates is an odd and inconspicuous one, not clearly destined for success. By good fortune not preparatory design, the later invaders of the land had fin structures arranged in such a way that the weight could be supported on land but not for that purpose obviously – unless the future controls the present.

Take another example: most people think that mammals arose near the end of the reign of dinosaurs and by virtue of their inherent superiority, at least played some role in the death of dinosaurs. But that is not true. Mammals arose at about the same time as dinosaurs. Two thirds of the history of mammals is not the history of their success after the death of dinosaurs but the history of their life as little creatures living in the nooks and crannies of the dinosaurs' world. They were not displacing these creatures. Something happened 65 million years ago. Perhaps the dinosaurs were already in decline, perhaps they were not. You are well aware that one popular theory for what happened, and I think it is likely to be so, argues that at least the trigger of the extinction was the impact of an extra-terrestrial object – certainly the ultimate bolt from the blue, the unpredictable old deux ex machina. If that is the reason for the death of dinosaurs and if mammals got through because they were small, perhaps, with larger population sizes – yet small is not an act of adaptation for survival in cometary impact but a negative result of never being able to invade the dinosaur's world – then again we are here by good fortune. And that is the good luck of it.

To end this theme, I will read something I wrote at the end of an essay on this subject: 'You can ask in conclusion if all these reveries really make any difference; I mean who cares in the Old Spirit of America at its pragmatic best; it's kind of fun to imagine oneself as a sort of divine disc jockey sitting before the tape machine of time with a library of cassettes labelled "Priapulids", "Pogonophorans", "Polychaetes", "Primates". But would it really matter if all these reruns that I've talked about produce their unrealistic opposites?' After all, you can say, 'I don't care about invertebrates,' but I would suggest to you, as I have, that everything would be different. You see the divine tape player holds a million scenarios, each perfectly sensible. Little quirks at the outset, chosen for no particular reason seem inevitable in retrospect. But the slightest early nudge tracks a different groove and history veers into another plausible channel diverging continually from its original pathway. The end results would be so different; the initial perturbation so apparently trivial. If *Opabinia* has survived. If *Pikaia* the first chordate had died, I have no confidence that *Australopithecus* would ever have walked erect on the savannas of Africa. And so you see, for ourselves I think we really can only say, with only a slight emendation of Miranda: 'O brave *and improbable* new world that has such people in it.'

The individual in Darwin's world part two then: the more abstract issue of what individuality means. Now that seems pretty clear doesn't it? We know what an individual is in Darwinian theory or in our ordinary language. Right, individuals are bodies, separated entities, each of us termed an individual. Below we are made of parts. Above we aggregate into collectivities which we call societies or countries or whatever. I would like to suggest to you that this is a culture-bound concept, that it is a clearly and definitely historically bound concept, and that it is central within Darwinian theory, and it *is* central in Darwinian theory, because Darwin's theory is about natural selection working on these kinds of individuals, on bodies. Darwinian theory is explicitly about the struggle of individual organisms, bodies, for reproductive success, and everything else flows from that. I would like to suggest that it is not a necessary factual statement about the world but a culturally and historically bound notion, and I would love the opportunity to tell that story because it is a story that does come right out of Edinburgh and I can now return to Adam Smith. If we ask why it is that Darwin chose to construct such a theory in which the actor, the agent of natural selection, should be bodies and only bodies, working for their individual reproductive success, the reasons are complex, but a very important part of the theme has to do with cultural embedding. People usually see the necessary cultural embedding of science as negative, as something that holds back the march of scientific progress. If we could free ourselves from the constraints of our culture, our politics, our psyche, we could move forward into that world of truth. It does not always work that way, although it happens frequently. Sometimes imbeddedness and culture can be very positive and my favourite example of that is the way in which Darwin could draw upon the laissez-faire system of Adam Smith to construct his theory. Darwin's theory does proclaim that the individual is the body and only the body, made up of parts below, aggregating into collectivities above, but here the agent of the process is only selection among bodies. The main reason Darwin constructed so uncompromisingly brilliant and definite a theory – it is anything but imprecise – is largely, I think, that he wished to overturn the world in which he had grown up. As a philosophical revolutionary he had wished to overturn the world of Paley's *Natural Theology*. Paley's *Natural Theology* maintained that from the good design of organisms and the harmony of ecosystems you could infer not only the existence of God but His goodness, His nature, His benevolence, His plans, His purposes. And they were inherent, all those attributes of God, in the higher order construction of the world of nature – that is in the harmony of ecosystems and the good design of organisms. Darwin set out to overturn that. If you wish to overturn that, how do you do it? What's the most radical argument in principle? I can imagine two arguments of which Darwin chose the more radical. The first argument might be: well, there are animals that are well designed out there and there are harmonious ecosystems but you know, you start looking in detail and there is a lot of imperfection out there. You know there are these wasps that paralyse caterpillars, and lay their eggs in the living flesh of the caterpillar. The young are born, and eat the living paralysed caterpillar from the inside. Maybe good for the wasp but not good for the caterpillar. There are a lot of things happening out there that might make you want to reassess the notion that the benevolence and good order of the world indicates God's intrinsic nature. Now that would have been a fairly radical argument – but it was not Darwin's argument. Darwin chose a more radical argument. The more radical argument accepts the premise that, yes, most organisms are well designed, yes, ecosystems tend to be harmonious but – and here is the almost perverse turning of Paley on his head – that this is true, it is only a side consequence. There is no law or principle that operates directly for it. It is an accidental side result, a consequence of something else that is working at a totally different level, namely, the only thing happening out there is that individual organisms are struggling for personal reproductive success. That is all. It sounds ultimately selfish. And yet the good design of organisms and the harmony of ecosystems in Darwin's theory is nothing but a side consequence of something that has apparently an opposite moral force. It will not serve to validate Paley's God. All that is happening out there is that organisms are struggling for personal reproductive success and nothing else. Everything that Paley used to infer the nature of God is side consequence, that is all. Now, where did Darwin get that argument? And it is only recently with the deciphering and publication of all Darwin's notebooks that it is been recognised that probably the strongest component of the input into the theory of natural selection that Darwin developed in 1838 was his reading of Adam Smith's life by Dugald Stuart. Because it is from the laissez-faire economics of the Scottish economic school that the brilliance of that distinctive notion of individuality comes. That is what Darwin is doing. He is trying to grasp the notion of individuality that originated here. The paradox of Adam Smith's system is that you might think superficially that if you want the best ordered economy you should get all the smart economists, meet them around a table. The Lord Provost then gives them ultimate powers to make laws to regulate the economics of the City of Edinburgh and you pass those laws. In other words you have imposed it from above like Paley's God imposed it from above. But Adam Smith says, 'No, you want to do just the opposite'. In fact you want to do something that seems guaranteed to produce the opposite result! You want to let individuals struggle for profit, that is all. You let them do it, untrammelled laissez-faire. But if you let that happen then those who do it well drive out those that do not do it so well. Those who do it very well balance out each other, and you end up with a maximally ordered economy through the action, and here comes the greatest of all metaphors in the history of the field, through the action of the 'invisible hand'. And it is truly an 'invisible hand', there is no director for it, there is just individual struggle for personal profit. You get that maximally ordered economy by consequence of a process at a lower level – the struggle among individuals that seems to have an opposite force. So individuality in Darwin's theory comes out of this city in that sense. Now I would like to say that at least from my point of view it is a great irony of Adam Smith's system that it does not work in economics, and it cannot. We are moral agents, we cannot bear the incidental loss involved in letting laissez-faire work in an untrammelled way. There is too much death and destruction. But nature does not care. Nature is amoral, not immoral. If it takes thousands of deaths to produce something good in nature, then that is how it works. And so Adam Smith's system works in nature but does not work in society. So I think, in short, this is where it comes from, the distinctive notion of individuality in Darwin's theory. Again Darwin's theory is very narrowly focussed, not everyone appreciates that. Darwin's theory is not just that anything out there that is for the good of the species, the good of the organism, is preserved. Darwin's theory is not that at all. Darwin's theory says that organisms struggle for individual reproductive success and that everything comes from this. Natural selection works on bodies and individuals are bodies. That has been the argument, but does it hold? Is it right? And, I think there are three reasons why it is not right. The third, which is most interesting, has to do with a fundamental revision of evolutionary theory, although I will give you two other reasons first. I think we have to construe the nature of the word individual far more broadly than we have. We have, after all, refined our language before. We have learned to say human instead of man for example so that we can include all, not half, of us. And I think we have to do a similar thing to the word 'individual'. We have to expand that concept of individual, beyond its focus on bodies. I think that there are three main reasons for that. First the question of scale. We view bodies as the ultimate individual because we are bodies and we see things at our scale. For us bodies are individuals and societies are collectivities. I am not sure it would appear that way to creatures of other scale. I cannot help thinking, if I may choose a somewhat facetious analogy, but I think it works, if you remember that old film 'Fantastic Voyage', where a group of scientists are shrunk to microbial dimensions and injected into a human body – it was Raquel Welch's screen debut – there is a wonderful scene where she is attacked by a bunch of antibodies. I could not help thinking that they are in what we would call a body but when they are shrunk down, that body becomes their collectivity. It is their universe. And so it is fundamentally a question of scale. We would not, were we not at our scale, choose to call our bodies the ultimate individual. I do not want to engage in fancy philosphical discussion about the nature of individuality. Let us just take good old vernacular notions of individuality and what it means. I suppose in the

English vernacular, individuals are at least minimally separate and different. If you do not have a bunch of things that are separate with the possibility of being different one from the other, we are rather suspicious about calling them individuals. My first point was a question of scale. Now I would like to suggest to you as my second point – ambiguities in continua – that even if you accept this vernacular definition having to be separate, having to be different, that although most organisms are discrete there is enough ambiguity in nature that it is not always so clear. For example, on the theme of separations as a necessary criteria for individuality there are conjunctions. They are not common, but I do not know what you do with them. The phenomenon of Siamese twinning in human beings is an obvious example. I am not talking about people like Eng and Chang, the originals, who are separate individuals joined at a tiny little strand of flesh, I am talking about those rather more tragic cases of people who are in part two and in part one, and there is not much more you can say about them. There is the famous case of Rita and Christina who had two heads, two upper bodies joined at the waist and then one lower body. Are they two or are they one? If you want to use the culturally invented definition of heads then you might say they were two. But then what about the opposite cases, and they do exist, of two lower parts and one upper part? These are ambiguous cases, I do not think there is more you can say about it. Likewise, on the opposite theme of dismemberment, occasionally interesting ambiguities occur.

You know for example, many lizards can knock off their tails, a process called autotomy. Now I am not trying to suggest that a knocked off tail ought to be construed as an individual – after all they do not live very long. But there is an interesting study that was done with good control several years ago, indicating that at least in some species of lizard that knock off their tails there is direct natural selection working on the motion of the tail. Now it may be, so to speak, for the escape of the body but nonetheless there is identifiable selection working on the autotomised tail for its flexibility and length of motion. And at that point can you deny some notion of individuality for the tail if it is a separate and distinct object upon which natural selection can work?

And then we come to the second theme of difference. What do you do with those species like aphids where the single female produces young parthenogenetically, identical to her, for generations and generations? There will be a sexual generation eventually to reduce the ambiguity but a single female aphid going through the parthenogenetic cycle for many generations can produce millions of genetically identical individuals. Are they bodies? Or are they simply separated parts of her? And indeed, Dan Janzen, the famous American evolutionist, has proposed that this is an unresolvable ambiguity. Perhaps we ought not to call each individual aphid a whole body. Maybe the whole collectivity is what he calls an EI, or an evolutionary individual. It is ambiguous.

We then have even stronger ambiguities on the continuum of sociality. Let us consider that continuum of sociality, people to ants to plants to siphonophores. It is unambiguous for people. We have no troubles with people. We aggregate into collectivities. But each of us is different from everyone else and undoubtedly a personality unto himself. Now we get to ants and it gets a little more ambiguous. The worker ants are not genetically identical, they are not clones, and yet their behaviour is so stereotyped and so invariant, the degree of social cohesion and interaction is so great, that many people in the history of the study of ant society have been tempted to talk about the entire hive as a super-organism. And then you get to plants, like bamboo and dandelions, in which each stem coming out of the ground appears to be a plant, a whole individual plant if you will, and looks to be so, but all of them are joined underneath and each is, in a sense, just a stem off a common underground runner. What do you call these things? The case is so ambiguous that botanists have developed a new terminology as to what is the individual and what is the part. They talk about 'ramets' and 'genets'. The genet is the whole thing, it is the whole underground system and all the stalks. But a ramet is each individual branch. This new terminology does not solve anything. We cannot resolve it. Is the stalk of a bamboo a plant or is it a part? It looks like a plant and yet it has these connections. And then when you get to siphonophores, you are absolutely stuck. You cannot resolve it. Siphonophores are a group related to corals and jellyfish. They are technically colonies by history, that is each of the parts historically, in evolutionary terms, was a separate individual. And yet the parts are specialised and separate in function – you have some parts of the siphonophore that only make reproductive cells, others that only eat, others that only protect, others that are only a part of the flotation device. In the technical literature of siphonophores each of these individual parts is actually called a 'person'. Biologists talk about polyp persons and feeding persons and floating persons right in this technical literature – it is charming, because they cannot resolve it –inherent ambiguities truly exist.

So finally I come to the third, last and most important theme on the question of individuality namely, the locus of selection. It sounds abstract and theoretical but I suggest to you that it is important. What is it in fact upon which natural selection works? Was Darwin right that natural selection works only on bodies. I would like to suggest that this is not right and that probably the most interesting revision now occurring within evolutionary theory is the attempt by many to construct a hierarchical theory of natural selection in which natural selection is still the agency of change, but it does not only work on bodies, as in Darwin's theory, but works on a whole hierarchy of increasing levels in genealogy. It works on genes below and it can work on species above. That may be an unfamiliar notion but think about it. What does it take to be a Darwinian individual? I said you ought to try and reform your language – do not get rid of the term 'individual' but expand it beyond the definition that an individual can only be a discrete physical body. What does it take to be an individual within Darwinian theory? It takes five things, and I would like to suggest to you that other objects besides bodies have these five properties. You have a birth point. You have to have a death point. You have to have relative stability in between. There is nothing particularly Darwinian in that – I think anything you would call an individual would have to have those properties by definition – a birth point, a death point and stability. And then you have to have children because Darwinism is a theory of inheritance and genealogical passage. And your children have to be like you – there has to be a principle of inheritance. If lions give rise to elephants it would be chaos out there. Genes have those five properties. Species have those five properties. You are used to thinking of species as collectivities. Species evolve by branching off from a parental form. There may be a little ambiguity at the start but not for most of the duration of the species. Species have definite death points when they become extinct. Most species are more stable during the course of their geological lifetime than human beings are in the ontogenetic sequence of foetus to elderly individual. Species have children – they bud off and give rise to daughter species and their children are like them: that is the principle of inheritance. The logic of natural selection can work on species. It gives you a very different perspective: species are individuals too. Genes are individuals too. And we need this enlarged hierarchical theory: there is no preference for the body as an individual. There are so many elements within the hierarchy. Let me give you one example, it is a potentially very interesting case. It is widely acknowledged by evolutionists that if genes, many genes, did not occur in multiple copies, there would be no opportunity for the evolution of complexity. That is something we did not know when Watson and Crick developed the model of DNA in 1953. Many people thought then that the chromosomes carried a set of individual genes one after the other. We know that this is not true. Many genes occur in multiple copies – hundreds and thousands of them. And the obvious question was then asked, 'Why? Why do we get so many copies?'. Think of bacteria – they do not have multiple copies and they have been bacteria for millions of years. They are wonderful working machines but each gene has one copy, each gene does something necessary for the life of the bacterium; it is not available for fundamental transformation. You cannot transform it fundamentally into something else because you will lose the essential function performed by that gene. But if you have more than one copy, then one copy can continue to perform an essential function and the other is free to vary. So that redundancy of multiple copies is absolutely necessary to the evolution of complexity. And so you then come back to the question, 'Why do we have multiple copies?'. You say, well, I've just answered it. It is obvious you have multiple copies so that complexity can evolve. You would not have complexity without it. The minute you say that, I hope you realise the logical fallacy of this answer. That cannot be right unless our views on causality are totally awry. You cannot evolve something so that it will be useful for you ten million years down the road. Unless the future controls the present that cannot be right. The existence of the multiple copies may be absolutely necessary for the evolution of complexity, but that cannot be why they are there. So why are they there? One popular explanation relies upon gene level selection, that is why I bring it up. It is called the hypothesis of selfish DNA. (This has nothing to do with the notion of the selfish gene. The selfish gene is an attempt to validate the Darwinian world with narrow selection on low levels, whereas the selfish DNA is part of the hierarchical model.) The notion of selfish DNA states that some genes, called transposons are jumping genes, are able to make copies of themselves and insert these second and multiple copies within the genetic systems at various places, and that is the explanation for so-called redundant or multiple copy DNA. A Darwinian might thus reply: "Yes, I understand what you're saying but that's not an explanation – you just told me that some genes make copies and insert. That's not an explanation'. And then someone who supports selfish DNA says, 'But I gave you the explanation – some genes can make copies and insert'. The Darwinian says 'That's not an explanation'. You see to the Darwinian, explanations have to be framed in terms of what is good for bodies. It is not an explanation until you say why the multiple copies are good for bodies. This is a wonderful example of what we call cognitive dissonance because the selfish DNA exponent has provided a perfectly good gene level explanation – that some genes, acting as Darwinian creatures in their own world, have figured out how to play Darwin's game. Darwin's game is about making more copies of yourself in your own world. Some genes have learned how to do that. They make more copies of themselves in their world and that is all the explanation you need. They are their own agents at this lower level of selection. Now at some point if the selection is still operating in Darwin's world of bodies, you get too many copies – the energetic cost to bodies may get too large – and increase of copies may be suppressed by natural selection operating on bodies, which leads to the interest of this hierarchical theory, namely that all these multiple levels of selection interact with each other with negative feedbacks: gene level selection produces multiple copies; organism level selection suppressing it; organism level selection making organisms extremely specialised to local environments; species level selections knocking it back when environments change and organisms cannot adapt – there is a very interestingly different world out there. And so, if you accept this, you say, 'There is really nothing special about bodies as individuals'. They are legitimate individuals but there are many other levels of individuality, and selection upon them may turn out to be just as important for understanding the history of life as selection on Darwinian bodies.

Now, let me end with an ethical point – it is a very small ethical point by the way because I do not believe that one ever can go from a statement about the factual character of the world to ethical values about what is good – I hope you are not depressed at this point. First I said you were not ordained to evolve, now I say there is nothing special about your personhood. But just as I think there is an optimistic side to contingency – that is *Wonderful Life* – I will end with this ethical point which I hope is the optimistic side to the suppression of individualities as bodies only. And the ethical point is small because the facts of the world do not necessitate ethical values. But, let me put it this way: it has often been said that the hopes of necessary bonding among us, into higher level collectivities, have been rendered impossible by the truths of Darwin's world. Darwinism after all is said to be about individual struggle, individuals struggling for reproductive success. Therefore our hopes of this necessary brother and sisterhood in the larger sense, even though they might be ethically preferable, will be thwarted, frustrated inevitably by the Darwinian nature of the world, that world being the struggle amongst organisms themselves. But it is perfectly all right to say that in Darwin's world individuals always act in their own interests and the bald principle holds but it is no longer quite clear what an individual is. Groups and societies are individuals too. And I think therefore the growing ties among us, that universal brother and sisterhood to which we aspire, is perfectly good biology, and the enlarged notion of individuality is not just outdated liberal hope or mushy romanticism. Of course I hope we never abrogate our focus on bodies as well, if only for that ultimate of political reasons, namely, that we weep for the inhumanity of some systems that did not cherish the worth of individual bodies, but did flourish for a time nonetheless. Still, I would repeat in conclusion that it is not outdated liberalism or mushy romanticism but perfectly good biology to say, in that most famous statement from my own tradition, as Rabbi Hillel has said, 'If I am not for myself, then who shall be for me? But if I am for myself alone, then what am I?'.

APPENDIX I

INTRODUCTION BY ELEANOR McLAUGHLIN, LORD PROVOST

Citation

The Edinburgh Medal has been instituted by the City of Edinburgh District Council to honour men and women of science who have made a significant contribution to the understanding and well being of humanity. It is awarded in 1990 to Stephen J. Gould in recognition of his contribution to a better understanding of evolutionary theory, through his exploration of the richness and subtlety of life and humanity's place within it, and especially for his uncompromising efforts to bring to the lay public a genuine and critical understanding of the nature of science and its interplay with society.

Ladies and Gentlemen it is with great pleasure that I welcome here tonight, on behalf of the City of Edinburgh, Professor Stephen Jay Gould to receive The Edinburgh Medal.

Last year I used this opportunity to say a few words on the International Science Festival and I am going to do the same this year.

Our first Festival last year was for us a step in the dark. We were confident that what we were proposing was worthwhile – even necessary – but conviction is not always enough.

In the event we were more successful than we thought we might be.

In ten days we organised 120 events to which 75,000 people came. This year we have expanded to a fortnight and our programme runs to 367 events. By tonight – after only one week – as many people as last year will have participated in the festival already.

And – as last year – from every continent in the world – including Antarctica.

Praise and thanks are due therefore to the dedicated organisers of the festival and to our citizens who have responded to the challenge.

But the numbers are important – not just for themselves – but for what I think they can also tell us.

That is that there is a popular hunger for understanding, not an abstract understanding – but an understanding that allows for a contribution to be made and a control to be exercised.

Many of the participants in the festival will tell you that the combination of scientists, technologists and lay people – from many disciplines and walks of life – has lead to sparkling and stimulating debates with fresh views and understandings – the consequences of which reverberate for individuals and groups long after the event is over.

Nor is it only the deepening of collaboration and understanding that the festival is fostering but practical proposals are emerging also.

Some of you will recall Professor Salam's suggestion made in his Medal Address last year. This has been taken up with enthusiasm by Edinburgh and Heriot Watt Universities and is taking the form of an international centre for mathematics – with an emphasis on working with mathematicians from developing countries.

Or this week – with one of the recommendations arising from the New Edinburgh Conversations – for the establishment – in collaboration with Leningrad – of an environment-monitoring station in Edinburgh as part of a worldwide chain of ten such stations.

These are practical outcomes but as Professor Gould commented recently: 'Science is an integral part of culture. It is one of the glories of human intellectual tradition.'

The role of the festival is not only to celebrate that glory but to extend it and to make it common to all humanity.

It is with this in mind that the city council has established The Edinburgh Medal and Address. Its objective is to honour an individual not just for past achievement but to offer that person a platform within the international festival to address the world under the general theme of science and society.

Professor Gould's work has been – amongst many other things – in the field of evolutionary biology. Possibly no other field of scientific work has given such persuasive metaphors for social and political life. From 'survival of the fittest' to the 'selfish gene' its impact upon society has been profound. Professor Gould's development of the theory of evolution has been a major contribution but as important have been his writings. In these he has conveyed – with clarity and warmth – in a way that can be understood by all – the beauty and richness of life.

But he is no Pangloss – and has consistently reminded us that science is not an inexorable march to truth through the collection of objective information. But – as much as any part of society – reflects the social and political constraints of the times – which has often led to the existing social order being defended incorrectly as biologically determined. An essential reminder – within our common celebration – of the continuing need for critical enquiry.

Thus it is for Professor Gould's extension of our deeper understanding of the origin and development of life and the manner in which he has conveyed that knowledge to all of us – which has undoubtedly added to – 'the glory of the human intellectual tradition' – that this city honours him.

APPENDIX II

ORATION

T. Dalyell MP

At the tender age of five, Professor Gould records that his father took him to the American Museum of Natural History in Manhattan. There he saw the reconstruction of a dinosaur, a twenty foot high Tyrannosaurus. 'As we stood in front of the beast,' he recollects, 'a man sneezed. I gulped. I prepared to utter "Myshela Israel" but the great animal stood immobile in all its bony grandeur, and as we left I announced that I would be a paleontologist when I grew up,' though we might be excused for being just a little surprised that even he used the word 'paleontologist' at the age of five! Maybe, had it not been for that visit, his whole life would have followed a different course. Not only as a student at Colombia and now Professor of Geology at Harvard, but as Curator of Invertebrate Paleontology at the Museum of Comparative Zoology in Cambridge, Massachusetts he will appreciate more than most the value of exhibitions on young minds, and the value of Science Festivals on young minds, even if five is indecently precocious. In visiting the UK he comes to a country where all is not well. We have had museum charges and the appalling results are becoming clear. The Science Museum in South Kensington in 1987 had 3 116 000 visitors. After the imposition of charges, attendances for 1989 were 1 121 000 – slashed to almost a third. Again the great

National Railway Museum in York had charges rather earlier – similar results; 1 113 000 in 1985, 542 000 in 1989. He comes to a country which needs stimulus.

Because he is a stimulator of scientific curiosity Professor Gould is enormously welcome in the city of Edinburgh with its associations with Darwin and other evolutionary scientists about whom he has written so eloquently and of whom he is one in a distinguished tradition.

APPENDIX III

VOTE OF THANKS

Ian Rolfe

We are gathered in a room with a mural depicting Apollo with the Muses. Apollo is the embodiment of the classical Greek spirit, the rational and civilised side of Man's nature. That rationalism is the basis of science and of what is needed in today's society, and indeed to safeguard our planet. This rationality has been exemplified by our speaker tonight. Not only in his speaking tonight, of course, but by his very many writings which many of you here know. Not just by his writings but by his activities also, in such spectacular ways as his fight against the Creationists, (in their unreason we may call them Dionysian as the opponents of Apollo), in the 1981 Little Rock Hearings.

In the painting Apollo is crowned with the laurel as victor, just as tonight Stephen Gould has been honoured and crowned by the City of Edinburgh with its second Edinburgh Medal.

Now the nine Muses by which Apollo is surrounded include Urania, the Muse of Astronomy, which must be taken to represent science in those days. Fortunately the artist who dated that painting in 1822 had updated his imagery and has added later poets and philosophers such as Shakespeare, and indeed Robert Burns himself appears. But not only such luminaries from the world of the arts; we also see Napier, Newton and Bacon to epitomise more modern science. They are surrounded by springs within the grove. These springs represent the refreshing sources of creative inspiration and of learning, which apply equally of course to the world of the arts and the sciences.

Sir William Dawson, Principal of McGill University and a well-known geologist, made an interesting comment in 1892 in the light of his own experience of the famous Scottish geologist Lyell's visit to Canada. He made this statement, talking about himself as opposed to Lyell: 'Patient plodding may accumulate large stores of fact, but the magic wand of genius is required to bring out the true value and significance of these stores of knowledge'. That is so true of science and this is what Steve Gould has done for us in his writings, and I suggest in his talk here tonight. Thank you Professor Gould.

Published by
EDINBURGH UNIVERSITY
PRESS

on behalf of
THE CITY OF EDINBURGH
DISTRICT COUNCIL

in association with
THE EDINBURGH
INTERNATIONAL FESTIVAL OF
SCIENCE AND TECHNOLOGY

Design contribution by
TAYBURN DESIGN GROUP
EDINBURGH